COUNTDOWN TO SPACE

JUPITER—
THE FIFTH PLANET

Michael D. Cole

Series Advisors:
Marianne J. Dyson
Former NASA Flight Controller
and
Gregory L. Vogt, Ed. D.
NASA Aerospace Educational Specialist

Enslow Publishers, Inc.

40 Industrial Road	PO Box 38
Box 398	Aldershot
Berkeley Heights, NJ 07922	Hants GU12 6BP
USA	UK

http://www.enslow.com

Library of Congress Cataloging-in-Publication Data

Cole, Michael D.
    Jupiter—the fifth planet / Michael D. Cole.
        p. cm. — (Countdown to space)
    Includes bibliographical references and index.
    Summary: Describes the history, unique features, and exploration of
Jupiter, the fifth planet from the Sun.
    ISBN 0-7660-1511-4
    1. Jupiter (Planet)—Juvenile literature. [1. Jupiter (Planet)]
I. Title. II. Series.
QB661 .C683 2001
523.45—dc21
                                       00-009262

Printed in the United States of America

10 9 8 7 6 5 4 3 2 1

To Our Readers:
All Internet addresses in this book were active and appropriate when we
went to press. Any comments or suggestions can be sent by e-mail to
Comments@enslow.com or to the address on the back cover.

Illustration Credits: National Aeronautics and Space Administration
(NASA)

Cover Illustration: NASA (foreground); Raghvendra Sahai and John
Trauger (JPL), the WFPC2 science team, NASA, and AURA/STScI
(background).

Cover Photo: Jupiter with two of its moons—Io (far left) and Europa.

CONTENTS

Jupiter and four of its moons were photographed by the Voyager 1 *spacecraft in 1979. Io (upper left) is closest to Jupiter. Then come Europa (center), Ganymede, and Callisto (lower right). They are not to scale.*

1

Jupiter Up Close

Galileo was getting nearer its destination. The robotic spacecraft had already traveled on a course around the planet Venus and through the inner part of the solar system—more than 2 billion miles. It was at last approaching the world it was designed to study: the largest planet in the solar system, Jupiter.

Galileo had separated into two parts: the orbiter and the atmospheric probe. Back on Earth, many scientists and engineers waited for the important moment when *Galileo*'s probe would enter Jupiter's atmosphere.

"I wonder what *Galileo* will discover," said Steven Tyler, a scientist on the mission. "I suspect that the probe will find something surprising."[1]

The probe plunged into the planet's atmosphere right on time and successfully sent its data to the orbiter above. Then it was time for the engines on *Galileo's* orbiter to fire. The engine burn put the spacecraft into orbit around Jupiter. Days later, *Galileo* began sending data to Earth from the distant planet.

"I can't wait to find out what interesting things they have found!" said Leslie Tamppari, another scientist on the mission. "It's so great to finally be in orbit around Jupiter!" Tamppari, like many others involved with *Galileo*, had been waiting to get the spacecraft to Jupiter for six years. "I was nearly in tears from the joy of knowing that we had done it."[2]

The *Galileo* scientists had reason to be excited about their accomplishment. Previous spacecraft had simply flown by Jupiter, sending back data about the planet for only a brief time before passing onward into space. *Galileo* was the first spacecraft to go into orbit around the giant planet. An orbiting spacecraft gave the scientists an opportunity to study Jupiter over a number of years. Tamppari and the other scientists could not wait to get started.

The news had excited one scientist's mother, too. Mission scientist Claudia Alexander's mother heard the report about *Galileo* on the nightly news. She finally understood the years of work her daughter had put into the *Galileo* mission.

"She said, 'OHHHH. Is that what you do. . . . We're going to learn all these exciting things about Jupiter for the first time. . . . This is really exciting.'"[3]

2

Giant of the Solar System

Early astronomers, from at least as far back as the ancient Greeks, identified Jupiter as something special among the thousands of other points of light in the sky. It was one of the bright objects that moved slowly through the unchanging background of stars. Our view of the stars changes from month to month as Earth travels in its orbit around the Sun. But the stars remain in their same patterns, or constellations, year after year, century after century.

Jupiter and the other visible planets, however, move through the sky from one area of stars to another. The word *planet* comes from the Greek word meaning "wanderer." The Romans later named the bright planet after Jupiter, the king of the gods in their mythology.

Jupiter's Size and Orbit

Jupiter is by far the largest planet in our solar system. If our solar system were viewed from some distant star system through a telescope, Jupiter would likely be the only planet detected. The remaining eight planets in our solar system could fit inside the sphere of Jupiter with room to spare. While the diameter of Earth is nearly 8,000 miles (13,000 kilometers), Jupiter's diameter is more than 88,700 miles (143,000 kilometers). At that size, Jupiter could hold more than 1,200 planets the size of Earth.

These images of Jupiter (left) and Earth were taken at different times. Jupiter, the largest planet in our solar system, could hold more than 1,200 Earths. Jupiter's Great Red Spot is the red oval in the bottom half of the planet.

Jupiter is 318 times heavier than Earth.[1] The result of this tremendous mass is that the planet's gravity is more than two and a half times stronger than Earth's. If you weighed 80 pounds on Earth, you would weigh over 200 pounds in Jupiter's outer atmosphere.

Jupiter, the fifth planet from the Sun, orbits the Sun at a distance of about 484 million miles (781 million kilometers). When Jupiter and Earth are on opposite sides of their orbits, Jupiter can be more than 577 million miles (931 million kilometers) from Earth. But Jupiter is so huge that even at these distances, it can be easily seen from Earth with the naked eye.

It takes Jupiter almost twelve Earth years to complete one orbit around the Sun. If you were ten or eleven years old on Earth, you would still be waiting to experience your first birthday on Jupiter.

The Gas Giant

Jupiter is what astronomers call a gas giant. It has no solid surface like Earth. If astronauts could travel to Jupiter, they would not be going there to land on the planet. Since Jupiter is made of gas, there would be nothing to land on.

Jupiter is made mostly of hydrogen and helium gases, with traces of methane, ammonia, and other gases in its clouds. The clouds on Jupiter are not made of floating droplets of liquid or frozen water like the clouds on Earth. Jupiter's clouds are made of frozen bits of gases.

Jupiter is a gas planet. Seen here is not a solid surface, but bands of clouds that surround the planet. The cloud bands are thousands of miles thick. The dark spot is a shadow of one of the planet's moons.

These clouds encircle the planet in bands. The temperature in these cloud bands is extremely cold, about -240° F (-151° C). Scientists believe the cloud bands, which are visible through telescopes on Earth, are thousands of miles thick.[2]

Underneath the cloud layers, the planet is surrounded by a thick layer of liquid hydrogen. The liquid hydrogen layer is probably 18,000 to 20,000 miles (29,000 to 32,000 kilometers) deep. The temperature and pressure within the planet increase as you go deeper into it. Before reaching Jupiter's core, the temperature is in excess of 30,000° F (16,650° C). At the greater pressure, the hydrogen no longer acts like a gas. It acts like a molten metal. Scientists call this substance metallic liquid hydrogen.

Jupiter's Core

Jupiter's core, about 35,000 miles (56,000 kilometers) beneath the planet's clouds, is probably made of rock. It is unknown whether this core is solid or if it is partly molten. Either way, the core is certain to be very hot. Most of it is probably iron, with ammonia and methane making up the rest. The core is about 18,000 miles (29,000 kilometers) in diameter and has a crushing pressure nearly 100 million times greater than the atmospheric pressure at Earth's surface.[3]

Scientists believe the intense pressure at Jupiter's core causes the core to compress, or shrink, by about one inch every year. One inch of shrinkage on a planet that is so huge may not seem like very much. But the shrinkage from compression produces intense heat. As a result, Jupiter gives off nearly twice as much heat as it receives from the Sun.

Magnetic Field and Radiation

The heat moving outward from the core through the metallic liquid hydrogen layer causes the liquid to move in outward and inward cycles. This movement spreads the heat and causes the hydrogen to conduct electricity. The flow of electricity through this layer creates the planet's magnetic field. Earth and most of the other planets in our solar system have magnetic fields, but Jupiter's magnetic field is a thousand times more powerful than Earth's.[4]

Particles streaming from the Sun, called the solar wind, come in contact with Jupiter's magnetic field. The interaction produces belts of radiation around the planet. This radiation would be deadly to any human being exposed to it. Earth's atmosphere shields us from the harmful radiation that exists in space. Jupiter's intense radiation would cook our bodies many times faster than sunlight (another kind of radiation) tans our skin. It would also damage the instruments of any spacecraft that travels too close to the planet for extended periods.

Rotation

Although Jupiter is gigantic compared to Earth, a day on Jupiter is less than half as long as a day on Earth. It takes 23 hours and 56 minutes for Earth to rotate once on its axis. Jupiter rotates every 9 hours and 55 minutes. This rapid rotation, and the fact that the planet is made up almost entirely of gases, causes Jupiter to bulge out

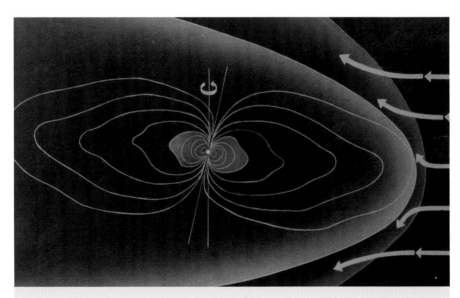

When particles from the Sun (yellow arrows) hit Jupiter's magnetic field (blue lines), radiation belts are produced.

farther in the region around its equator. As a result, the diameter of the planet is 5,700 miles wider from east to west than it is from north to south.[5] The rotation of the planet is also five minutes slower at the poles than at the equator.

Great Red Spot

Jupiter has a striped appearance when viewed from space because of the bands of clouds around the planet. Pictures from spacecraft reveal the complexity of the cloud bands. Their borders at times mix together in extremely detailed patterns. At other times they appear not to mix at all. The result is that the cloud bands display a rich variety of color and contrast. Some of the cloud bands move around the planet at speeds greater than 300 miles per hour.

Giant circular storms often swirl through the cloud bands. Some of the storms last ten years or more. The largest of these circular storms is called the Great Red Spot, a dominant feature of Jupiter's atmosphere. Astronomers have been observing the storm through telescopes since the 1660s, making it more than 340 years old. It is observable using telescopes because of its huge size. Three Earths could fit inside the 25,000-mile-wide storm.[6]

The Great Red Spot swirls in a counterclockwise motion. It takes about seven days for the storm to rotate once around its center. The entire storm moves around the planet in a westward direction. It never drifts to the north or the south. It is bounded on the south by a mild easterly wind and on the north by the strongest westward wind on the planet.

Amid Jupiter's clouds, there is lightning one hundred times more powerful than that on Earth. If such lightning were to strike the ground on Earth, it would likely cause a blast equal to that of hundreds of pounds of TNT—enough to level a four-story building.

Far beyond the clouds of Jupiter's upper atmosphere, a collection of smaller worlds orbits the giant planet. Numerous and different, they are the moons of Jupiter.

Jupiter's Great Red Spot, which is three times the size of Earth, is a giant storm rotating in Jupiter's atmosphere.

Jupiter's Great Red Spot with four of Jupiter's moons—Io, Europa, Ganymede, and Callisto. The Galileo spacecraft took the images of Jupiter, Io, Ganymede, and Europa. Callisto's image was taken by Voyager.

JUPITER

Age
Probably about 4 billion years

Diameter at equator
88,700 miles (143,000 kilometers; more than
100 times Earth's diameter)

Diameter at poles
82,950 miles (133,800 kilometers)

Planetary mass
318 Earth masses

Distance from the Sun
484 million miles (781 million kilometers; 5 times
Earth's distance from Sun)

Closest passage to Earth
391 million miles (631 million kilometers)

Farthest passage to Earth
577 million miles (931 million kilometers)

Orbital period (year)
11 Earth years, 316 days

Rotation period (1 day)
9 hours and 55 minutes (slower at the poles)

Temperature
-240° F (-151° C) in outer atmosphere

Composition
Solid or molten rock core (scientists unsure), layer of metallic liquid hydrogen, vast outer layer of liquid hydrogen, mostly hydrogen gas atmosphere

Atmospheric composition
86% hydrogen; 14% helium; traces of methane, ammonia, and water vapor

Wind speeds
380 miles per hour maximum at the equator

Gravity
About 2 1/2 times Earth's gravity
(in Jupiter's outer atmosphere)

Number of moons
17

Composition of rings
Dust, which was blown off small inner moons by meteor impacts

Diameter of Great Red Spot
25,000 miles (wider than three Earths)

Amount of heat received from Sun
4 percent of the amount Earth receives

3

Many Moons

In 1610, Italian astronomer Galileo Galilei became the first person to use a telescope to make detailed observations of Jupiter. Jupiter was the first planet at which he aimed his newest invention to study.[1] He discovered something quite unexpected. Four objects appeared to be moving very slowly around the planet. Galileo had discovered moons in orbit around Jupiter. These moons, named Europa, Ganymede, Io, and Callisto, are today called the Galilean moons, after their discoverer.

At least seventeen moons orbit Jupiter. They vary in size and composition. Most of them are less than one hundred miles in diameter and are irregularly shaped. The Galilean moons are by far the largest.

Ganymede

Ganymede is covered with ice and is more than 3,000 miles (4,800 kilometers) wide. That is larger than Earth's Moon, which is 2,160 miles (3,480 kilometers) wide. Ganymede is not only the largest moon of Jupiter, but the largest moon in the solar system. It is even larger than the planets Mercury and Pluto. If Ganymede orbited the Sun instead of Jupiter, it would easily be classified as a planet.

Ganymede is made almost entirely of ice and rock. Parts of it are dark brown or dark green, while other areas are light brown or light red. The darker surfaces have many craters. Ganymede's lighter colored terrain has complex patterns of cracks and grooves.

The smoother, grooved terrain was probably caused by the release of water from beneath the moon's surface. The water froze and later cracked. These icy regions may have been formed and reformed many times, making this terrain a much younger surface than the darker regions. The darker, rougher surfaces have many more craters than the smooth areas. This means that space objects have been hitting

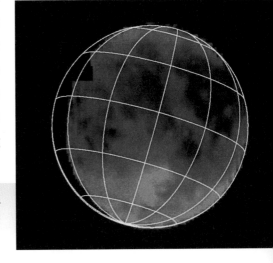

This false-color image shows water ice on Ganymede's surface. The brighter areas have more ice.

these rougher regions for millions of years longer than the smooth areas. Some scientists believe the darker, cratered terrain is the original crust of Ganymede.[2]

Europa

The smallest of the Galilean moons is Europa. It is also covered with ice that is cracked and grooved. But unlike the brown, red, and dark green colors of Ganymede, the ice on Europa is smooth and white. These qualities make Europa the most reflective and brightest of Jupiter's moons. It is the smoothest known object in the solar system. It is five times brighter than Earth's Moon.

Many substances besides water can be frozen into ice form, but evidence from *Galileo* so far indicates that the ice covering Europa is made of water. Some scientists believe that an ocean of water may exist beneath Europa's icy surface. If such an ocean exists, it is possible that some forms of life as we know it could exist within it.[3]

Io

Io is the Galilean moon that orbits nearest Jupiter. It is about the size of Earth's Moon but looks nothing like it. Active volcanoes cover Io. Its surface is coated with sulfur blown out from previous eruptions. This volcanic activity gives Io an unusual appearance from space. It is often called the pizza moon, because the red-and-yellow coloration from the sulfur reminds people of the tomato sauce and cheese of a pizza.[4]

Some of the eruptions on Io are so massive that they

can be sighted through telescopes from Earth. One volcano, called Loki, is so enormous that its lava spreads over an area larger than the state of Maryland. Loki is the most powerful volcano in the solar system, putting out more heat than all the volcanoes on Earth combined. Some volcanoes on Io produce heat greater than 1,000° F (538° C), and the hottest one there is recorded to have temperatures of over 3,000° F (1,650° C). This means that in some places, Io has the hottest known surface in the solar system. Its volcanoes produce a hundred times more lava than all of Earth's volcanoes put together.[5]

The tiny particles blown into space by the volcanoes cause something bizarre to occur around Io. As the moon moves in its orbit through Jupiter's magnetic field, electrons in the magnetic field strike the particles around Io, causing

Europa is the smallest Galilean moon. It is covered with smooth white ice. The top image is a natural color. The color of the bottom image has been enhanced.

About the size of Earth's Moon, Io is covered with active volcanoes.

them to glow. The result is that whenever Io passes into Jupiter's shadow, a thin ring of glowing gas outlines the moon. The electrons moving along the planet's magnetic field lines also cause an electric current to flow between Jupiter and Io. The flow of electricity across the 217,000 miles that separate them makes the largest electrical circuit in the solar system.[6]

If you stood on Io and looked into the sky, you would feel as if you were standing almost nose to nose with Jupiter. Io orbits so close to Jupiter that the giant planet would literally fill the sky.

Callisto

Callisto is the darkest Galilean moon and orbits at the greatest distance from Jupiter. It orbits at more than one million miles away from the planet—more than four times the distance between Earth and its Moon. Callisto's surface is made of ice and is dark gray. The many white splotches on its surface are the sites of impact craters. Scientists believe that the impacts caused water to rise to the surface from beneath, forming newer ice.

Callisto is the most heavily cratered object yet observed in the solar system. The number and size of craters suggest that Callisto is more than 4 billion years old.[7]

Two of the planet's smaller moons orbit less than 80,000 miles (129,000 kilometers) outside its atmosphere. Four others orbit at a distance of more than 13 million miles (21 million kilometers). The most recently discovered moon of Jupiter was found in 2000.[8]

Although detailed study of Jupiter and its moons began with Galileo and his telescope in 1610, much of our knowledge about the planet comes from the spacecraft that have traveled there over the last few decades. The development of spacecraft that could cross the hundreds of millions of miles to Jupiter began in the late 1950s.

The bright spots on this image of Callisto are impact craters. Meteorites have exposed the lighter material below the moon's darker surface.

4

Spacecraft at Jupiter

The Space Age began in 1957 when the Russian space program launched the first satellite, *Sputnik*, into orbit around Earth. Then the space programs of both the United States and Russia began to explore the solar system by sending spacecraft to the Moon and to the planets Venus and Mars.

Early Planetary Spacecraft

NASA's *Pioneer 10* was launched in March 1972. It was the first spacecraft ever to travel beyond Mars. On December 3, 1973, *Pioneer 10* flew by Jupiter, becoming the first spacecraft to reach the giant planet. While making its way past Jupiter, *Pioneer 10* sent twenty-three pictures of Jupiter's atmosphere back to Earth.

In 1974, *Pioneer 11* flew by Jupiter, taking seventeen

more pictures of the planet. It sent back data on the temperature and pressure within the planet's atmosphere and took a number of pictures of Jupiter's moons. *Pioneer 11* then used Jupiter's gravity to change course and continue to Saturn.

Voyager Missions

The missions of *Voyager 1* and *Voyager 2* were far more ambitious than the Pioneer missions. These two spacecraft, which flew by Jupiter in March and July of 1979, returned 30,000 images of the planet and its moons. They took pictures of the highly complex cloud patterns in Jupiter's atmosphere. The detailed images forced scientists to totally rethink their theories about the causes of wind and weather patterns on the planet.

The spacecraft also detected a faint ring system around Jupiter. The rings are nothing like the complex system of rings around Saturn. Saturn's rings are made of billions of bits of ice. Ice reflects light very well, making the rings of Saturn easily visible with telescopes on Earth. Jupiter's rings are made of dust particles. The dust particles get thrown into space by meteorites impacting the surface of the small moons orbiting nearest Jupiter. The debris makes a trail of dust along the orbital paths of the small moons, forming a ring around Jupiter. But since dust does not reflect light as well as ice, Jupiter's rings can only be seen with very powerful telescopes.

The Voyager spacecraft that flew by Jupiter in 1979 helped scientists better understand the planet's clouds and weather patterns.

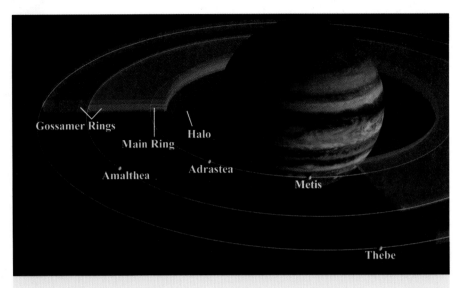

The rings of Jupiter are formed by dust from the small moons that orbit close to the planet.

Voyager 1 was the first to photograph a cloud of gas and dust erupting from an active volcano on Io. Scientists had previously believed that all moons were frozen hunks of rock or ice. *Voyager 1*'s photograph was the first evidence scientists had seen of volcanic activity on any moon in the solar system.[1]

Six weeks after *Voyager 1* ended its encounter with Jupiter, *Voyager 2* began making observations. *Voyager 2* took pictures of Callisto and Ganymede during its journey toward the planet. High-quality images of Europa were obtained as well, revealing its smooth, ice-covered surface with its many cracks and fractures but few craters.[2]

Voyager 1 visited Saturn before heading out of the

solar system. *Voyager 2* also flew by Saturn, then continued to Uranus and Neptune. To make a study of Jupiter and its moons in much greater detail, scientists needed to place a spacecraft in orbit around the planet. Named after the discoverer of Jupiter's moons, that spacecraft was *Galileo*.

Galileo Spacecraft Journeys to Jupiter

Galileo began its long journey to Jupiter after astronauts launched it from the space shuttle *Atlantis* in October 1989. The spacecraft's planned flight path carried it through the inner part of the solar system before traveling out toward Jupiter.

In 1994, when the spacecraft was still 121 million miles from Jupiter, *Galileo* took pictures as fragments of the comet Shoemaker-Levy 9 smashed into Jupiter's atmosphere. A comet is a body of ice and dust that travels around the Sun in elliptical, or oval, orbits. When a comet gets nearer the Sun, the ice in the comet is turned into vapor and forms a long tail behind the comet. Comet Shoemaker-Levy 9 had been broken apart into more than twenty fragments when the comet's orbit carried it too close to the gravitational pull of Jupiter.

The images amazed astronomers. After viewing the first few images from the Hubble Space Telescope, scientists saw that the first comet fragment had caused a gigantic disturbance in Jupiter's atmosphere.

"We realized that we had something truly spectacular on our hands," said one scientist.[3]

By observing the comet impacts, scientists could learn a great deal about comets and about Jupiter's atmosphere. For example, they could take spectrometer readings, which is a way of analyzing light, during the fragments' collisions with Jupiter. Studying the light emitted from these collisions would tell astronomers what specific elements made up the comet.

While the event's importance to science was obvious,

Galileo *is ready to be launched from the space shuttle* Atlantis *and begin its journey to Jupiter.*

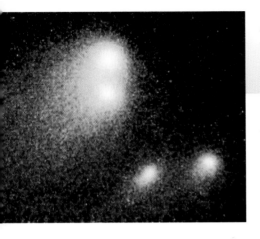

In 1994, comet Shoemaker-Levy 9 caused a huge disturbance in Jupiter's atmosphere. The Hubble Space Telescope captured the moment.

one reporter at a press conference wanted to know how important it would be to the public.

"What's in this for the guy in the street?" asked the reporter.

"The guy in the street," one scientist replied, "can be glad he doesn't live on Jupiter!"[4]

Galileo Arrives at Jupiter

Galileo continued its journey toward the planet. With 147 days until the spacecraft's arrival, *Galileo*'s atmospheric probe separated from the orbiter. Both spacecraft continued on separate paths toward Jupiter. They arrived at the planet on schedule, on December 7, 1995. It was a busy day for scientists and engineers at NASA's Jet Propulsion Laboratory (JPL) in Pasadena, California. The JPL is the control center for *Galileo* and other unmanned spacecraft. The probe was about to enter Jupiter's atmosphere, and the orbiter's engine had to fire successfully to slow the orbiter down and put it into orbit around the planet. Everything had to work perfectly.

"The problems of probe entry and survival . . . and of

the relay link are significant," said engineer Steven Tyler.[5] If the probe entered Jupiter's atmosphere too quickly, it would burn up. If its parachute did not deploy, the spacecraft would drop through the atmosphere too quickly for its instruments to do their job. If the probe's transmitter had been damaged during the long flight, none of the data the instruments collected would be relayed to the orbiter. Any malfunction with the spacecraft meant that no information about the planet's upper atmosphere would be gained.

Tyler and other JPL workers watched anxiously at their stations. Although they had no pictures yet, the information on their monitors told them that the probe was descending on a parachute through Jupiter's clouds. As the probe went deeper and deeper through the clouds, it relayed data about Jupiter's atmosphere to the orbiter passing overhead. The orbiter stored the data, which it would later relay to Earth. After

The assembled Galileo *probe was photographed prior to its mission.*

about an hour, the probe was crushed, as expected, by the mounting pressure within the giant planet. Its data about Jupiter's atmosphere was already safely stored on computer tape aboard the orbiter.

The next important step was firing *Galileo*'s engines to put it in orbit around the planet. Controllers at JPL waited for the time when the engines would start their burn.

"We got the signal that the big engines aboard *Galileo* started firing right on time," said JPL scientist Leslie Tamppari. "That was a relief too since if they didn't work right, we would have become a flyby mission instead of an orbiting mission!"[6]

With the engine burn successfully completed, *Galileo* became the first spacecraft to go into orbit around Jupiter. After a few days, scientists at JPL received the relayed data from the atmospheric probe.

The probe's mission into Jupiter's clouds was a huge success. Its instruments recorded wind speeds of several hundred miles per hour in the planet's cloud bands and revealed the elements that make up Jupiter's upper atmosphere. The amount of helium gas detected by the probe told scientists that the deep interior of Jupiter might be an ocean of liquid helium.[7]

Probe scientist Richard Young was pleased that the spacecraft had revealed some new and surprising things about the giant planet.

"The probe also discovered an intense new radiation

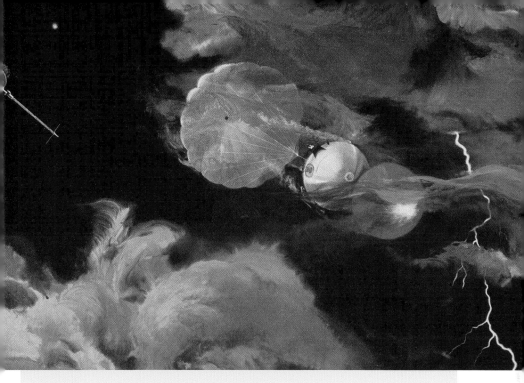

An artist created this image of the probe parachuting through Jupiter's atmosphere. The probe successfully sent data about the planet's atmosphere back to Earth.

belt approximately thirty-one thousand miles above Jupiter's cloud tops." he said."[8]

Jupiter's four largest moons were a source of many surprises for *Galileo* scientists. Images from the spacecraft showed that sulfur dioxide blown from active volcanoes had turned some areas of Io's surface white. Ash and lava from volcanic eruptions had noticeably changed the appearance of Io's surface during the seventeen years since the *Voyager* missions. Its appearance continued to change during the *Galileo* mission.

"Some of the lavas erupting on Io are extremely hot, as high as 3,140° F (1,727° C)," said planetary geologist Alfred McEwen. "That's much hotter than present-day

A volcanic plume on Jupiter's moon Io rises about 60 miles (100 kilometers). Some of the lava on Io is much hotter than lava erupting on Earth.

eruptions on Earth, but similar to lava that commonly erupted on Earth in the early Precambrian period."[9] The Precambrian period was a period in Earth's distant past, before 570 million years ago to as much as 4.6 billion years ago.

"So when we look at Io," added project scientist Torrence Johnson, "we're looking at a place that is in some ways a lot like Earth was long ago."[10]

Galileo also discovered a magnetic field around the largest of Jupiter's moons, Ganymede. It is the first moon ever discovered to possess a magnetic field. A thin atmosphere of hydrogen and carbon dioxide was detected around Callisto. What appeared to be frozen streams of water ice were recorded on the icy surface of Europa, giving further evidence that liquid water may exist below Europa's surface.

Galileo Continues on Extended Mission

In 1997, at the end of the original two-year mission, the *Galileo* science team extended the spacecraft's mission by three years. The secondary mission produced important images of Jupiter's White Ovals. These are the many long-lived circular storms, smaller than the Great Red Spot, that move through the planet's cloud bands. In 1996 and 1997, during the initial mission, the spacecraft showed two White Ovals getting very close to each other.

"It became clear they would merge," said Torrence Johnson. "And when we imaged them again in September 1998, the two White Ovals had become one. We caught one of the first major changes in Jupiter's atmosphere in more than sixty years."[11]

The extended mission also produced some interesting discoveries on Callisto. Images from the *Voyager* missions had scientists expecting to see craters of every size on Callisto's icy surface.

"But what we saw," said *Galileo* scientist Ronald Greeley, "was an eroded surface with very few craters smaller than a half mile. The surface is being eaten away and blanketed by soft, fluffy stuff."

Images of Callisto showed craters blanketed with fine debris. Greeley believed that ice was playing a key role in

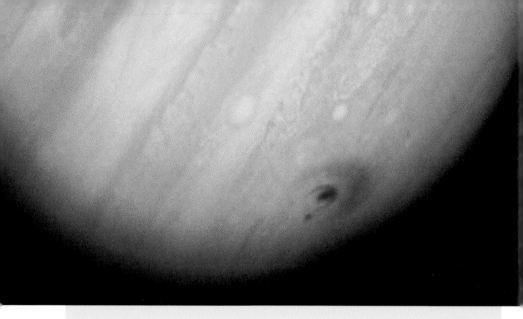

The Hubble Space Telescope captured this image of Jupiter. Continued collection of information from space missions will help scientists learn more about the planets, our solar system, and the universe.

producing the "fluffy stuff" that smooths Callisto's surface.

"As the ice [evaporates] and is lost, all that is left is the dirt," Greeley said. "It's like after your snowman melts, and all that's left is that little pile of soil."[12]

Galileo's mission at Jupiter will continue until the planet's radiation finally causes the spacecraft's instruments to fail. Even after the signal from *Galileo* is lost, the spacecraft will remain in orbit around Jupiter or one of its moons for many years. Gravity will eventually pull the spacecraft from its orbit, causing it to crash on the surface of a moon or plunge into the crushing pressures of Jupiter's atmosphere.

Data from *Galileo* and other spacecraft, as well as images recorded by the Hubble Space Telescope, have given scientists a new and deeper understanding of Jupiter. Such knowledge helps scientists answer important questions about how planets are formed at different distances from the Sun. It also gives them a better picture of what the early solar system was like.

Astronomers using new telescopes and light-measuring equipment have identified planets in orbit around a number of other stars. The planets themselves have not been seen. They have been detected by measuring the slight "wobble" of a star, caused by the gravitational effects of orbiting planets. Planets are also detected when they pass directly in front of their star, causing the amount of light reaching Earth from the star to decrease.

So far, the light measurements have told astronomers that most of these planets are very massive, like Jupiter. Knowledge about our own solar system's largest planet has aided scientists in understanding what these planets around other stars might be like.[13]

Jupiter was once as mysterious and unknown to us as the planets around other stars are today. Because of telescopes and spacecraft, we no longer have to imagine what Jupiter is like.

5

Observing Jupiter from Earth

For now, the best and easiest way for humans to observe Jupiter is from Earth. Observing Jupiter is a treat. The planet's size and banded clouds make it one of the most interesting planets to watch through binoculars or a telescope.

Jupiter appears far larger in a telescope than any other planet. Even binoculars will reveal the four largest moons. With binoculars or a telescope, you can watch the movement of these moons. Each night, the position of the moons will be different. Monthly astronomy magazines such as *Sky & Telescope* and *Astronomy*, as well as some newspapers, offer charts that show you which of the moons you are seeing in each position for every night of that month.

Another interesting activity is watching the shadow of one of the moons move across the planet. Depending on which moon is casting the shadow, its journey can be observed from start to finish within a few hours. If the shadow is moving very slowly, it is the shadow of Callisto. Callisto, the most distant Galilean moon, takes sixteen days and seven hours to orbit the planet.

An average-sized telescope will also show the Great Red Spot as it moves along its path around the planet. The Great Red Spot currently looks pale orange through a telescope, and it has looked that way for a number of years.[1]

Some of Jupiter's moons, which you can see from Earth with a telescope, are shown in size comparison with other celestial bodies.

If binoculars or a telescope is unavailable, see if there is an astronomy club in your area. Many communities have active astronomy clubs and programs for the public. Most clubs work through public libraries, park systems, or museums. By attending these programs, you will learn about astronomy and be able to look through telescopes at planets such as Jupiter and many other objects in space.

Jupiter, with its immense size and many moons, is almost like a mini solar system. Scientists believe that if Jupiter had been ten times larger, its interior pressure would have caused a collapse. Such a collapse would have resulted in nuclear reactions that would have turned the more massive Jupiter into a star like the Sun. Because of this, some scientists think of Jupiter as a failed star.[2]

Jupiter, in reality, is a giant planet with colorful banded clouds, orbited by a collection of uniquely interesting moons. Visible to the naked eye from a distance of more than 500 million miles, Jupiter is truly the giant of the solar system.

CHAPTER NOTES

Chapter 1. Jupiter Up Close

1. NASA K–12 Educational Resource, "Steven Tyler Field Journal Entry," *Online from Jupiter*, December 7, 1995, <http://quest.arc.nasa.gov/galileo/bios/fjournals/tyler-ofj3.html> (September 12, 1999).

2. NASA K–12 Educational Resource, "Leslie Tamppari Field Journal Entry," *Online from Jupiter*, December 11, 1995, <http://quest.arc.nasa.gov/galileo/bios/fjournals/tamppari-ofj2.html> (September 12, 1999).

3. NASA K–12 Educational Resource, "Claudia Alexander Field Journal Entry," *Online from Jupiter*, December 11, 1995, <http://quest.arc.nasa.gov/galileo/bios/fjournals/alexander-ofj3.html> (September 12, 1999).

Chapter 2. Giant of the Solar System

1. Reta Beebe, *Jupiter: The Giant Planet* (Washington, D.C.: Smithsonian Institution Press, 1994), pp. 1–2.

2. Jean Audouze and Guy Israel, eds., *The Cambridge Atlas of Astronomy* (Cambridge, England: Cambridge University Press, 1996), pp. 164–166.

3. Ibid.

4. J. Kelly Beatty, Carolyn Collins Petersen and Andrew Chaikin, eds., *The New Solar System* (Cambridge, Mass.: Sky Publishing Corporation, 1999), pp. 196–197.

5. Audouze and Israel, p. 166.

6. Michael E. Bakich, *The Cambridge Planetary Handbook* (New York: Cambridge University Press, 2000), p. 213.

Jupiter Table

J. Kelly Beatty, Carolyn Collins Petersen, and Andrew Chaikin, eds., *The New Solar System* (Cambridge, Mass.: Sky Publishing Corporation, 1999), pp. 193–219; Jean Audouze and Guy Israel, eds., *The Cambridge Atlas of Astronomy* (Cambridge, England: Cambridge University Press, 1996), pp. 164–189.

Chapter 3. Many Moons

1. William R. Newcott, "In the Court of King Jupiter," *National Geographic*, September 1999, p. 130.

2. "Ganymede Fact Sheet," *Project Galileo Home Page*, n.d., <http://galileo.ivv.nasa.gov/ganymede/> (September 16, 1999).

3. "A Summary of Facts About Europa," *Project Galileo Home Page*, n.d., <http://galileo.ivv.nasa.gov/europa/e-summary.html> (September 16, 1999).

4. Torrence V. Johnson, "Jupiter and its Moons," *Scientific American*, February 2000, p. 45.

5. Paul Recer, "Study Finds Even More Volcanoes on Jupiter Than Thought," *Honolulu Advertiser*, November 19, 1999, p. A9; J. Kelly Beatty, Carolyn Collins Petersen and Andrew Chaikin, eds., *The New Solar System* (Cambridge, Mass.: Sky Publishing Corporation, 1999), p. 10.

6. Johnson, pp. 46–47.

7. Jean Audouze and Guy Israel, eds., *The Cambridge Atlas of Astronomy* (Cambridge, England: Cambridge University Press, 1996), pp. 177–179.

8. Reuters, "Astronomers Find Possible 17th Jupiter Moon," July 22, 2000, <http://www.cnn.com/2000/TECH/space/07/22/space.jupiter.moon.reut/index.html> (August 1, 2000.)

Chapter 4. Spacecraft at Jupiter

1. Jean Audouze and Guy Israel, eds., *The Cambridge Atlas of Astronomy* (Cambridge, England: Cambridge University Press, 1996), pp. 186–187.

2. Reta Beebe, *Jupiter: The Giant Planet* (Washington, D.C.: Smithsonian Institution Press, 1994), pp. 18–19.

3. David H. Levy, *Impact Jupiter: The Crash of Comet Shoemaker-Levy 9* (New York: Plenum Press, 1995), p. 154.

4. Ibid., p. 156.

5. NASA K–12 Educational Resource, "Steven Tyler Field Journal Entry," *Online from Jupiter*, December 7, 1995, <http://quest.arc.nasa.gov/galileo/bios/fjournals/tyler-ofj3.html> (September 12, 1999).

6. NASA K–12 Educational Resource, "Leslie Tamppari Field Journal Entry," *Online from Jupiter*, December 11, 1995, <http://quest.arc.nasa.gov/galileo/bios/fjournals/tamppari-ofj2.html> (September 12, 1999).

7. Torrence V. Johnson, "Jupiter and its Moons," *Scientific American*, February 2000, pp. 45–46.

8. NASA Press Release 96–10, *Galileo Probe Suggests Planetary Science Reappraisal*, January 22, 1996.

9. Ibid.

10. William R. Newcott, "In the Court of King Jupiter," *National Geographic*, September 1999, p. 134.

11. Ibid., p. 130.

12. Ibid., p. 139.

13. Geoff Marcy and Paul Butler, "Hunting Planets Beyond," *Astronomy*, March 2000, pp. 43–47.

Chapter 5. Observing Jupiter from Earth

1. Phil Harrington, "Falling for Jupiter and Saturn," *Astronomy*, October 1999, p. 92.

2. J. Kelly Beatty, Carolyn Collins Petersen, and Andrew Chaikin, eds., *The New Solar System* (Cambridge, Mass.: Sky Publishing Corporation, 1999), pp. 194–195.

GLOSSARY

atmosphere—The layers of gases surrounding an object in space.

cloud bands—Cloud formations, such as those around Jupiter, in which the clouds are separated into many individual horizontal areas that each completely encircle the planet.

comet—A celestial body that travels in a huge elliptical orbit. When orbiting near the Sun, it develops a long tail that points away from the Sun.

constellation—A pattern or arrangement of stars in a given area of the sky. There are 88 recognized constellations, each with its own name, such as Orion and Leo.

crater—A violently disturbed area created by the impact of another object from space.

flyby mission—A mission in which a spacecraft makes its observations as it passes a planet or other object in space. Many early space missions were designed to fly by, not to orbit or land on, the object they were sent to study.

Hubble Space Telescope—An orbiting observatory equipped with a very powerful telescope, designed to view objects up to 13 billion light-years away.

magnetic field—The area around a star, planet, or moon where forces due to the electrical current within that body can be detected.

NASA—National Aeronautics and Space Administration; the United States government agency in charge of the country's space program.

planet—Any of the nine large bodies in our solar system that orbit the Sun.

radiation—The emission of energy in the form of waves or particles. High levels of radiation can be deadly to humans.

ring—Billions of bits of dust or ice that form an orbiting trail around a planet.

satellite—Any object in orbit around another object. Spacecraft in orbit around Earth are called satellites. The Moon orbits Earth, making it a natural satellite of Earth.

solar system—The Sun, its planets, and their moons.

solar wind—The flow of charged particles emitted by the Sun that extends throughout the solar system, far beyond the orbit of Pluto.

space shuttle—The reusable, winged vehicle used by NASA for transporting astronauts, experiments, satellites, and spacecraft into space.

sulfur—A nonmetal element that exists on Earth, on other planets, and on some moons, such as Io.

volcano—A vent in a planet's or moon's crust from which molten material is ejected from the interior.

FURTHER READING

Books

Beebe, Reta. *Jupiter: The Giant Planet.* Washington, D.C.: Smithsonian Institution Press, 1994.

Berger, Melvin. *Discovering Jupiter: The Amazing Collision in Space.* New York: Scholastic, Inc., 1995.

Cole, Michael D. *Galileo Spacecraft: Mission to Jupiter.* Springfield, N.J.: Enslow Publishers, Inc., 1999.

Landau, Elaine. *Jupiter.* Danbury, Conn.: Franklin Watts, Inc., 1999.

Rogers, John H. *The Giant Planet Jupiter.* New York: Cambridge University Press, 1995.

Internet Addresses

Arnett, Bill. "Jupiter." *The Nine Planets.* June 22, 2000. <http://seds.lpl.arizona.edu/nineplanets/nineplanets/jupiter.html>.

Hamilton, Calvin J. "Jupiter." *Views of the Solar System.* © 1997–2000. <http://www.solarviews.com/eng/jupiter.htm>.

Lowes, Leslie. *Galileo Home Page.* "Galileo: Journey to Jupiter." July 6, 2000. <http://galileo.jpl.nasa.gov/>.

The Regents of the University of Michigan. *Windows to the Universe.* © 2000. <http://windows.engin.umich.edu>.

Scientific American. *By Jove!* n.d. <http://www.sciam.com/exhibit/030397jupiter/030397galileo.html>.

INDEX